KIT
REPAIR

Printed in the United Kingdom by MPG Books, Bodmin

Published by Sanctuary Publishing Limited, Sanctuary House, 45-53 Sinclair Road,
London W14 0NS, United Kingdom

www.sanctuarypublishing.com

ISBN: 1-86074-384-6

basic KIT REPAIR

ROBBIE GLADWELL

Also in this series:

basic CHORDS FOR GUITAR
basic DIGITAL RECORDING
basic EFFECTS AND PROCESSORS
basic GUITAR WORKOUT
basic HOME STUDIO DESIGN
basic LIVE SOUND
basic MASTERING
basic MICROPHONES
basic MIDI
basic MIXERS
basic MIXING TECHNIQUES
basic MULTITRACKING
basic SCALES FOR GUITAR
basic VST EFFECTS
basic VST INSTRUMENTS

contents

chapter 4

chapter 5

introduction

Keeping the instruments, equipment and other essential paraphernalia a busy on-the-road musician inevitably hauls around in tip-top condition will not only enhance the music-making process by ensuring trouble-free rehearsals and gigs but may well save you from a red face when you come to play that all-important venue. At best, faulty leads, amplifiers or pedals may mean your much-practised solo goes unheard; at worst, it may mean pure electricity running through your fingers!

With a bit of thought and only a few basic tools, you can maintain your equipment so as to prevent problems or breakdowns both in rehearsal and on-stage. *basic KIT REPAIR* gives practical advice on the issues musicians are confronted with on a daily basis, ranging from choosing an instrument case, shipping equipment, re-stringing guitars and troubleshooting electrical problems to maintaining PA systems and amplifiers, cleaning cymbals and changing drum heads. Remember, you don't need to be an expert to keep your kit in good order. In fact, if any problem you come across is beyond your capability, you'll glean enough information from this book to recognise when it's serious enough to require professional attention.

With regard to basic maintenance, you'll need a good safe working area, but not necessarily anything as professional as a workbench. A simple table and tool kit is the most you'll need. You probably have most of the tools you'll need, anyway, and those that you don't can be purchased relatively cheaply from your local DIY store.

Safety issues feature heavily in this book, especially with regard to electrical items. A lot of the problems people come to me with could easily have been rectified by the owners of the instruments – which is the point of this book.

If you are pointed in the right direction, you can tackle simple repairs and basic maintenance and save yourself money in the process. It doesn't matter whether you're an amateur or a professional, your equipment can break down at any time if it's not correctly maintained.

chapter 1

careful with that axe!

Guitarists are notorious for not looking after their instruments. It seems that many of them believe that a battered guitar covered in filth and with the paintwork peeling off is the most desirable thing with which to be seen on-stage . This seems to be true irrespective of the age or playing style of the musician in question.

Conversely, there are those of us who are proud to be seen with a clean, shiny guitar. Obviously, the more you look after your guitar, the less likely it is to fail you during a gig. In truth, though, the cosmetic condition isn't that important with regard to functionality, but you should ensure that the fretting, pick-ups and electrics are cleaned and properly maintained. Okay, so while it may be cool to be seen with a beaten-up guitar, it certainly isn't cool if your audience is left waiting while you try to fix it in the middle of your set! A guitar should age gracefully and, if looked after, will achieve that aged look without becoming unreliable.

cases and covers

Many instruments are supplied complete with either a soft cover or a hard case nowadays. While these may be suitable for everyday storage at home, they may not survive the rigours of life on the road. This is especially true of soft covers and gig bags, which can be easily crushed by heavy amps or cabs should you swerve to avoid a badger while frantically trying to get to the gig on time!

Even some hard cases aren't actually as tough as they might look, and you should be very careful if you're travelling in a van with piles of heavy gear in the back. Make sure that the guitars are placed on top of the amps and cabs so that they can't fall or be crushed by other equipment.

If you do just the occasional gig, you probably needn't worry too much about upgrading your case – with a little care it should survive – but if you gig regularly and rely on other people to help with the equipment, then it pays to buy either one of the new light-weight "liteflite"-style cases or a full flightcase. These are usually quite expensive, sometimes as much as a reasonable entry-level guitar, so buying one makes sense only if you wish to protect an expensive – or cherished – instrument.

Heavy-duty flightcases are reinforced with either aluminium or modern plastics bonded to high-quality plywood. A light flightcase, on the other hand, will suffice for a medium-to-budget instrument, and these models are generally moulded from ABS plastic and lined with cellular polyurethane foam, adding to the case's rigidity as well as giving thermal protection to the instrument inside.

Guitar cases come in a wide variety of shapes, sizes and materials

It's essential that you buy a case that fits the instrument properly, as poorly fitting cases can cause severe damage to an instrument should the case fall over or be dropped. I've received any number of guitars in my workshop with broken necks caused by poorly fitting cases.

A heavily padded gig bag is fine if you intend to travel in your own car or on public transport and should protect the instrument adequately, as long as you don't drop it. If you're transporting your instrument in extremes of weather, a hard case is essential to protect it from thermal shock, which can cause severe damage to your temperature-sensitive guitar. Never leave an instrument unattended in the back of a car – or, for that matter, in any vehicle – for any length of time; apart from the security risks, on a hot day an instrument can sustain permanent damage even in the highest quality case.

Maintaining an instrument at a constant humidity can be very difficult. If it gets too dry, the timber will crack or split and the action height may increase. Conversely, if it gets too damp, the tone may not be as clear and the action height may also be adversely affected. Some acoustic-instrument specialists stock devices which, when placed in the case, help maintain the correct humidity for your instrument. Even electric guitars can be damaged by too much humidity, which causes the hardware to tarnish and the neck to warp.

On the other hand, if an instrument gets too dry, the fingerboard may split. To avoid suffering damage from sudden changes in humidity, keep your guitar in its case whenever possible.

If you've purchased a new or second-hand instrument without a case, your local music dealer should be able to source something suitable to protect it. If not, check on the Internet. There are manufacturers which specialise in a variety of cases that will fit even the most unusually shaped instruments, so there's no need for your cherished six-string to be without that vital protection!

transportation and shipping

Should you wish to send your guitar via a carrier – for set-up or repair, for example – make sure that it is adequately packaged before shipping. Sending it in its case alone does *not* provide sufficient protection! An alarmingly high percentage of guitars and cases get damaged in the hands of shipping companies, and proving that it's a company's fault can be difficult.

The best packaging method is to place the guitar in its case (with the strings de-tuned) and then place it in a large, strong cardboard box (your local music dealer may be able to provide a suitable container). To prevent the case from moving around inside the box, use rolled-up sheets of newspaper as packaging. Make sure that you've filled the space around the case, then tape the lid of the box securely with proprietary parcel tape. Label the box clearly on both sides with the destination address as well as your own address, making sure that your address is smaller than the destination address and using the words "to" and "from").

Acoustic instruments, in particular, suffer badly from extremes of temperature and humidity. A sudden change in either can cause cracks in the lacquer or, in the worst cases, actually in the timber. To help avoid this, always transport acoustic instruments in a proper case. For an electric guitar, you should at least use a padded gig bag.

If the instrument has been out in conditions below freezing, allow it to rest in its case for an hour or so to acclimatise before opening it. If this isn't possible for any reason, open the case lid for a few seconds and then close it for a minute or so. This should be done several times in order to reduce the likelihood of damage occurring to the instrument.

import and export

If you intend to travel abroad with your guitar or import/export a new instrument, there are various legal issues that you should observe if you want to avoid encountering problems with the authorities, either at home or in the country of destination when you arrive there. If you're a UK resident and wish to take an instrument to a country which is not a member of the European Union (EU), you would be wise to obtain a *carnet* before attempting it. This is a legal document which is generally obtained from the port of embarkation, or alternatively from your local Customs and Excise office in advance. A carnet allows you to temporarily export your guitar without having to pay the prevailing duty and local taxes at the destination country, although a small refundable bond may be payable upon entry.

If you're travelling within the EU and are a UK resident, then it's wise to carry proof of ownership – something as simple as a purchase receipt is fine. If you buy a guitar in another country, such as the USA, you must declare it at Customs when you return to the UK and present proof of purchase. This also applies if you're a US resident and you wish to export a guitar from the UK. You will be liable for import duty and local sales tax at the current rates.

It's important to remember that levels of taxation often change, as do local rules and regulations, so it's wise to consult the Customs and Excise authorities in both countries before you travel. If you buy any gear or instruments on the Internet from abroad, you'll have to arrange for it to be shipped by a company that specialises in exportation. They will arrange all of the necessary paperwork and will, of course, bill you for duty and import tax.

storage

If you need to store your guitar for any length of time, it's wise to take precautions. Firstly, de-tune all of the strings by a tone or more so as to prevent any possibility of the neck warping or pulling forward, should there be a fluctuation in temperature. Secondly, make sure that the instrument is cleaned and polished before you place it in its case. It's also a good idea to put a small bag of silica gel in the case in order to absorb any moisture build-up due to any fluctuations in humidity that may occur. Finally, place the case in the largest plastic bag you can find (a large dustbin bag will suffice) and store it in a cool, dry place out of direct sunlight and away from sources of heat or moisture. Suitable storage locations include a cupboard under the stairs (as long as it isn't damp), a dry cupboard in your lounge or bedroom, or under your bed.

The following places are not suitable for long-term storage of a guitar: garages, attics, kitchens, bathrooms, cellars, garden sheds or anywhere where there may be fluctuations in temperature or humidity. Make sure that you check on the instrument's condition at least two or three times a year to catch potential problems before they become serious.

stands

It doesn't matter whether you're an amateur, a semi-professional or a professional playing regularly, there will be occasions when you need to put down your guitar for one reason or another. Most players simply lean their guitar against an amplifier or chair, which is an accident waiting to happen! In my years as a professional repairer, I've seen countless broken guitar necks guitars, pretty much all due to the instrument falling over because the owner didn't have a guitar stand. Also, be careful not to leave trailing leads plugged in – they are easily tripped over!

Guitar stands come in various shapes and sizes, some of which are good and others of which are almost as bad as having no stand at all! The bad ones tend to resemble deckchairs in that they're cumbersome and difficult to put up, and they also tend to be rather top-heavy. For this reason, anyone passing by is at risk of knocking them over, and this can can cause terrible damage to a guitar.

In my experience, the simple stands are the best kind as they support the body of the guitar. The ones to avoid are those that hang the guitar by the neck. The simple ones fold out like an inverted V and the guitar body sits in place on two fold-out retainers.

Even if you play at home and don't perform live, it's essential that you have a stand tucked away

A good-quality stand is a must for every guitarist

somewhere safe on which to place your guitar. If your guitar has a cellulose finish, you must be very careful not to leave it on a stand for long periods of time. This is because most stands have rubber strips on the metal parts that come into contact with the guitar. Unfortunately, rubber reacts with nitro-cellulose paint and in severe cases can mark the finish or cause the lacquer to peel off.

One way around this problem is to cover the rubber areas with plain cotton or silk, or simply to drape a small tablecloth over the stand. Other types of finish do not need such protection, but unfortunately cellulose, although it looks good, is vulnerable to attack from sweat and synthetic substances. I know it's obvious, but I'll say it anyway: if there are young children in the house, it's probably wise to keep your guitar in its case and place it on its stand only when you're in the same room.

straps...

Another good investment is a decent guitar strap. There are many varieties available but, like everything else, you get what you pay for. I personally prefer a good leather strap, as it's less likely to mark a delicate finish such as those found on expensive guitars.

Plastic straps and the cheap woven ones are fine for use at home. Also, be aware that large buckles and studs can cause irrevocable damage to the finish of your pride and joy! It's also a good idea to purchase a set of Straploks at the same time.

...and straploks

The second greatest cause of broken necks is guitars falling off straps! Straploks are designed to prevent exactly this sort of disaster. They're normally quite easy to fit to most guitars and just screw in in place of your existing strap buttons. Straploks are made by several manufacturers but they all function in much the same way. I can guarantee that this is one purchase that won't disappoint you.

Straploks will ensure that your guitar stays on your shoulder

chapter 2

guitar and bass maintenance

Assuming that you have a working knowledge of the major components of a guitar, you shouldn't find these instructions difficult to follow. Study each section carefully, as it's important that you understand the techniques involved before you start work.

work area

Unless you have access to a ready-made workshop you will need to improvise, but fortunately guitar set-up needs minimal space. The basic requirements are: a clean, dry area roughly four feet by two feet at a comfortable working height and a power supply. Kitchen worktops are two feet deep and three feet from the floor, which is a comfortable height to work at if you're standing up. If you have a clear four-foot run with a mains socket behind, you may well find that this is the best place to work. Alternatively, try a small table, which is generally much lower and more suited to sitting at.

basic equipment

Before placing your guitar on the worktop, it needs to be supported at the neck, under the nut. This removes pressure from that area and leaves the fingerboard level. A neck support can be made from a piece of softwood, in the form of a block with a hollow cut out of it into which the neck can rest. This hollow should be lined with felt, or a similar material, to prevent damage to the finish. A piece of soft material, ideally felt or an old blanket, should also be placed under the body where it touches the worktop.

tool requirements

Before starting work, it's advisable to gather together all the tools that you think you might need, as this will save frustration and delays later. The following tools should be considered as minimum requirements. You will no doubt add other suitable tools and gadgets as you become more proficient.

tool list

- A smooth, flat, "single-cut" mill file
- A set of needle files
- Large and small cross-head screwdrivers
- $^1/_4$" and $^1/_8$" flat-blade screwdrivers

The typical guitar repairer's armoury

- A junior hacksaw blade
- Stanley-knife blades
- A small dovetail or gentleman's hacksaw
- A small steel rule graduated in 64ths of an inch
- A truss rod adjuster that fits your guitar
- 1,200-grade wet-or-dry paper
- 360-grade silicone carbide paper
- 000-grade wire wool
- A tube of Super Glue
- An adjustable spanner
- A small ball-pein hammer
- A set of Allen keys to fit the tremolo and locknut

assessment

Perform the following procedures to assess your instrument. The information required to make these adjustments and assessments can be found later in the book under the relevant headings. Take your time, and if you find anything that you're uncertain of, consult a professional repairer.

Tune to pitch A=440Hz and check the condition of the strings. If any of them are damaged or worn, replace them with your usual gauges. Finally, ensure that the strings are securely fixed to the machine-head posts in order to prevent slippage. Step by step, then, here we go:

1 Check the action height at the 12th fret
2 Assess the neck relief with a visual check from the headstock, down the neck to the body, and by fretting

You'll need a good selection of both cross-head and flat-blade screwdrivers

the first string at the first and last fret simultaneously and measuring the relief at the seventh fret

3. Make truss-rod adjustments, if required, and re-tune
4. Ensure that the pick-ups are lowered at this stage to allow an unobstructed view
5. Check the nut height at the first fret
6. Set the action height and re-tune
7. Trim the truss rod again and re-tune
8. Test-play a chromatic scale along each string to identify problems such as fret buzz or string cancellation. Identify these areas with a small cross on either side of the fret(s) using a soft pencil
9. If your instrument suffers from fret buzz, consult a professional in order to ascertain whether a fret dress is required. The pencil marks on the fingerboard will help the repairer to identify the problem areas
10. Check the guitar's intonation
11. Adjust the pick-ups to the correct height
12. Set up the tremolo (if fitted) by adjusting the spring tension

paintwork

You must be extremely careful when cleaning your guitar. Many cleaners and solvents can affect the paintwork and either mark or damage it. If your guitar is fitted with gold-plated hardware, avoid using any abrasive polish, as this will remove the plating. If the finish is covered in tiny scratches and minor belt marks, these can be removed by using a cutting polish such as T-Cut. This is a mild abrasive, however, so you should be very careful as you can easily cut through the finish! Apply T-Cut with a soft cloth using a circular motion. The best way of removing greasy marks is with a wax-free polish such as Mr Sheen. If your guitar has a cellulose finish, however, you should be particularly careful as it can be damaged by solvents found in some polishing compounds.

Also, be cautious when using thinners and paints. Minor chips can be touched up with car paint or suitably coloured nail varnish. Apply several light coats and let it dry for several days. You can use 1,200-grade wet-or-dry paper to cut the finish back to the level of the existing finish and then polish out the sanding lines using T-Cut. Avoid leaving your guitar in sunlight, as this can cause the finish to fade and crack. If possible, leave it in its case when you're not using it so that it doesn't become damaged by the environment or through an accident.

truss rod adjustment

A certain amount of neck relief is required to allow clearance for string vibration. Neck relief can be assessed in two ways:

Adjusting the truss rod

1 By sighting down the neck from the headstock along the line of the two E strings. A slight bowing of the neck should be apparent on both sides. A hollow relief will appear concave in relation to the strings, whereas a crown will appear convex.

2 By pressing the top or bottom E string down at the first and last fret, a measurement can be taken from the seventh fret to the underside of the string. It should be no more than 0.005" (use a feeler gauge) for steel-string instruments and no more than 0.010" for basses and nylon-string instruments.

Truss-rod adjustment will be necessary if relief is outside these parameters. By turning the truss-rod nut in a clockwise direction, the neck will take on a convex relief; counter-clockwise produces a hollow, or concave, relief. Under no circumstances should a truss rod be adjusted more than two complete turns, and you should always make adjustments with the instrument tuned to pitch. Generally speaking, a quarter- to a half-turn is sufficient in most cases. As a guide, if you require more relief at the seventh fret, slacken the truss rod slightly; if you require less relief, tighten the rod slightly. Finally, play a chromatic scale on each string from the nut to the last fret, listening for buzzes and stopping to make adjustments to the rod and action as necessary.

action height adjustment

Getting the correct action height is crucial to playability, but not as tricky as it might initially seem. We'll look at electric guitars first: measure the action height on the first and sixth strings at the 12th fret with a rule graduated in 64ths of an inch. Measurements should be taken from the top of the fret to the underside of the string. Guitars with individually adjustable bridge saddles require action measurement on each string. The optimum action heights for various styles of playing are as follows:

- **Rock And Blues Players** – Typical string gauges are 009 to 042; action height $^{3}/_{64}$ top E, $^{5}/_{64}$ bottom E. Players with a heavier technique using gauges

010 to 046; action height $^{3}/_{64}$ top E, $^{6}/_{64}$ bottom E. This will help to eliminate fret buzz.

- **Acoustic Steel-String Players** – Styles fall into two main categories:

1 Fingerstyle and solo players are advised to use a light-gauge string, typically 011 to 052; action height $^{3}/_{64}$ top E, $^{6}/_{64}$ bottom E

2 Plectrum and thumb-pick players are advised to use medium- to light-gauge strings, typically 013 to 056 with an action height of $^{4}/_{64}$ and $^{7}/_{64}$

Now set your action height according to your playing style. Les Paul-style guitars will have overall height adjustment either side of the bridge while Strat-style guitars will have individual height adjustment with provision for neck tilt, should the saddles be set to extremes. Many modern guitars have facilities for overall, as well as individual, height adjustment

If your guitar has individually adjustable bridge saddles, you'll need to measure the action on each string

acoustic guitars

Acoustic-guitar bridges play the dual role of anchoring the strings and transmitting string vibrations through to the table. The size, height and position of the bridge will depend upon the instrument's scale length, neck angle and string spacing.

On steel-strung guitars, the bridges are usually made of either rosewood or ebony. Occasionally, they can be made from materials that are "ebonised", meaning that they're painted black to hide the material's true nature! The bridge saddles are made from many materials nowadays, some man-made and some natural. Popular materials include hard plastics, synthetic bone and natural materials such as animal bone and ivory.

The saddle slots are angled to allow for intonation. On some instruments, there are two saddles, one for the E, A, D and G strings and the other for the B and E strings, and these are placed at different angles to allow for correct intonation on the top two strings.

Action height depends on personal preference, string gauge and playing style. Adjusting the saddle height on acoustic guitars is obviously a bit more involved than on an electric guitar. If the action on yours is too high, you should remove the saddle and place it in a small vice. Then, using a medium-cut file, remove a small amount of material from the base of the saddle, re-install it and check the action height again. Repeat the process until you achieve the desired action. If you think that you might have removed too much of the saddle in the process, it may be a good idea to consult a professional guitar repairer. If the action on your acoustic is too low (rare, but it does happen), you can place thin shims of wood veneer under the saddle. Simply cut the veneer to size with a pair of scissors.

Classical bridges are simpler in construction and feature a straight saddle with no apparent compensation for intonation. They are generally made from either rosewood or ebony but, as with some steel-strung bridges, lower quality materials are often used which are, again, ebonised. The strings are tied to the back of the bridge after being passed through small holes behind the saddle.

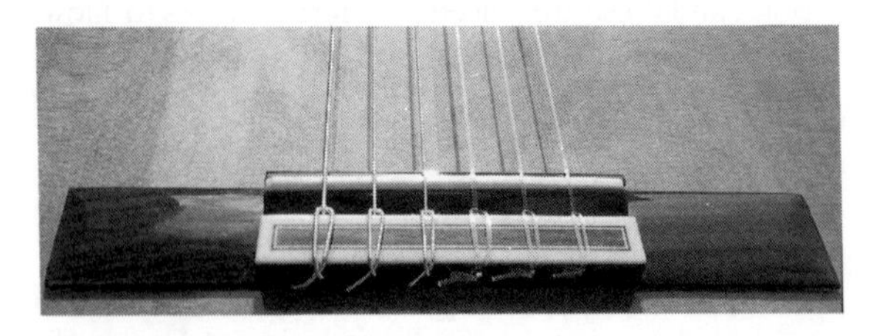

A typical classical-guitar bridge

On expensive models, the bridge saddles are generally fashioned from bone, but on cheaper instruments plastic is the material of choice. Adjusting the action height is pretty much the same as it is for a steel-string acoustic guitar. Don't be tempted to lower the action too much or the volume and tone will diminish considerably.

Classical guitar players are advised to use either medium- or high-tension strings, depending on their suitability to the instrument. Typical action heights are: $^{6}/_{64}$ top E and $^{8}/_{64}$ bottom E or higher (measured at the 12th fret).

On a general note, rarely do you see a bridge with a saddle which is adjustable for height or intonation on an acoustic. This is because any mechanism involved would have an adverse effect on the tone of the instrument. Adjustments to these instruments have to be carried out by a professional guitar repairer

intonation adjustment

For electric instruments, the bridge saddles are used to make adjustments to intonation. The object is to achieve equally tempered tuning right along the length of the fingerboard. Reference points are taken by playing harmonics at the 12th and 19th frets and audibly comparing them with the stopped notes at those positions. If the stopped note is sharp in relation to the harmonic, the saddle must be moved backwards, increasing the string length and so flattening the note. Conversely, if the note is flat in relation to the harmonic, the saddle must be moved forwards.

Adjusting the bridge saddles on an electric guitar

A simple tip is this: sharp = backwards, flat = forwards. For acoustic instruments, minor intonation problems can be corrected by feathering the saddle in the right direction. More serious intonation adjustments require re-positioning the bridge slot. It's a good idea to purchase an electronic tuner as this will allow you to make a more accurate assessment of the intonation.

fingerboard nut adjustment

Nut height is measured at the first fret in 64ths of an inch and can be broken down into four main string-type (gauge) categories:

- Light: E and A strings $^{2}/_{64}$ or lower; D and G strings $^{1}/_{64}$ or lower; B and E slightly lower than $^{1}/_{64}$
- Medium: E and A strings $^{2}/_{64}$; D and G strings $^{2}/_{64}$ or lower; B and E strings $^{2}/_{64}$
- Bass guitar: E and A strings $^{3}/_{64}$ or lower; D and G strings $^{2}/_{64}$ or lower
- Classical guitar: E and A strings $^{3}/_{64}$; D and G strings $^{2}/_{64}$; B and E strings $^{2}/_{64}$ or lower

Before making adjustments to the nut, make sure that the action height and truss-rod settings are correct, then slacken the strings slightly and pull them to one side so that they're clear of the slots. The slots can then be cut deeper with a needle file or a blade with a width close to that of the string.

Cut the slots a little at a time and check their first-fret clearance at regular intervals. If you cut the slots too deep you may require a new nut, as filling the slots isn't very successful as a repair. Incidentally, a little bit of pencil lead in the slots will prevent the strings from sticking and improve tuning stability.

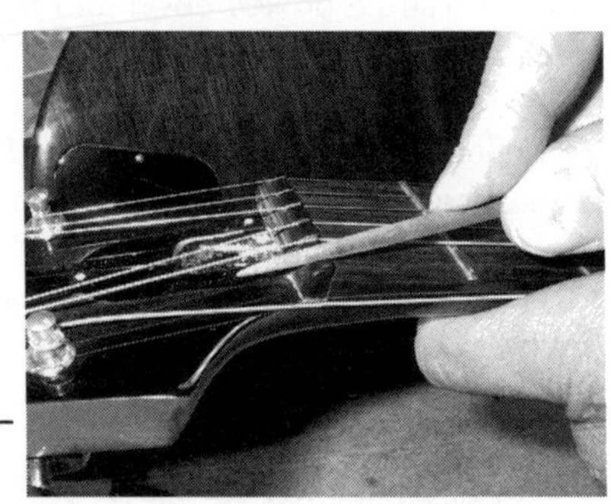

Nut height can be altered with the careful use of a file – take care not to overdo it!

frets and fingerboards

Over a period of time, the condition of the frets and fingerboard can deteriorate to the point at which either minor or, in some circumstances, major renovation is required. Providing that the frets aren't too worn and the fingerboard isn't unduly marked, you can perform this work yourself. Light fretwear can be rectified by sanding the frets with a fine-grade paper.

To do this, you should first remove the strings and inspect the fingerboard and fretting. If you notice any damage that's beyond your capability to rectify, or if the frets are too low, you should leave it and consult a professional. Otherwise, prepare your work area and assemble the items and tools required to perform the work.

Sand the frets lightly with 360-gauge silicone carbide paper using a sanding block. (It's imperative that you don't overdo this, as you could remove more height than is necessary.) Finish with 1,000-grade wet-or-dry paper (dry) and then go over the fingerboard with 000-grade wire wool to polish the frets.

Use a clean two-inch paintbrush to remove sanding dust and wire wool from the fingerboard and body of the guitar. While doing this, you should wear a face mask so that you don't breathe in any dust or debris. It's also a good idea to apply some fingerboard conditioner or lemon oil to the board.

Finally, re-string the guitar and perform a playing test. If you notice any undue fret buzz after this procedure, you should initially check the action and re-adjust the truss rod, if necessary. If this doesn't rectify the fret buzz you should consult a repairer to determine whether the problem is of a more serious nature.

strings

The strings you choose for your electric guitar will depend mainly on your playing style. If you play modern rock or blues, then you'll probably prefer the feel of lighter-gauge strings (with a plain third string). This is especially true if you do a lot of string bending.

If you are a rhythm guitarist or play old-style rock 'n' roll, you'll probably be better off with medium-light-gauge strings (with a wound third string). There are many different makes and gauges available, so it's really a case of experimenting until you find a weight with which you're comfortable.

Most strings are nickel-plated, so if you have a problem with nickel sensitivity, seek out an alternative. You can find strings that are plated with either gold or chrome that manufacturers often claim have a longer shelf life, although they're usually more expensive. The range of string gauges for steel-string acoustic guitars is also enormous. Before the advent of amplification, many acoustic guitars were fitted with heavy-gauge strings to aid projection of tone and volume. The tops of these instruments were often more heavily braced than they are today. Typically, a heavy-gauge set may range from 0.013 for the top E to 0.062 for the bottom E.

Most manufacturers recommend medium or medium-light gauges. Medium-gauge strings typically range from 0.012 to 0.056, whereas medium-light-gauge strings normally range from 0.011 to 0.052 (from the top E to the bottom E). Light-gauge or ultra-light-gauge strings are available for acoustic instruments, typically ranging from 0.010 to 0.047.

Of course, the gauges quoted here are typical and not necessarily those represented by any specific manufacturer. Check the manufacturer's website to get the details of particular sets of strings.

If you play bluesy styles that involve a lot of string bending, I would suggest you use light or ultra-light gauges. You may sacrifice a small amount of volume and tone, but bending will be much easier. For finger-pickers and folk guitarists, either light or medium-light gauges are best, while if you're looking for projection and volume, you should fit medium-gauge strings.

Unless the manufacturer of your particular instrument states otherwise, heavy gauges should be avoided due to the possibility of damage occurring to the modern, lightly braced tops. Acoustic guitar strings are made from various bronze alloys, with plain metal used for the B and E strings.

Again, manufacturers sometimes plate the strings with gold in an effort to increase the life of the string. This works fairly well if you're not string bending or strumming fiercely, but if you do thrash your guitar then metal fatigue will occur irrespective of the type or quality of the plating.

The strings found on classical guitars are generally made from nylon or, in the case of the lower strings, a multi-fibre nylon core with a metal wrap – usually a brass, bronze or silver alloy. Classical-guitar strings are not normally rated by their thickness but rather by their tension.

Beginners often use low-tension strings, whereas proficient players use medium- or high-tension strings for a firmer tone. The action height on a classical guitar is also set much higher than on a steel-strung guitar because the vibration pattern is much greater.

re-stringing acoustics

Stringing a guitar correctly makes all the difference to its tuning stability. A regular acoustic guitar is strung in a similar fashion to an electric guitar, with respect to how the strings are attached to the machine heads. An easy method is to tie a simple knot as you pass the string through the machine-head post so that the string locks onto itself as you tune up. This method ensures better tuning stability than merely relying on several layers of string around the post.

Tying a simple knot in the string will help ensure tuning stability

Certain vintage-style steel-strung instruments, such as those based on the early Martin and Washburn guitars, have "slot-head", side-mounted machine heads that are similar to those found on classical guitars. These are more fiddly to string up than conventionally mounted machine heads. However, there is a method that will make the job easier. In essence, it's just the stringing sequence that prevents you from tying knots in your fingers!

Starting with the two E strings, pass the strings through the holes in the metal machine-head posts, pulling them through so that only a small amount of string slack is evident over the fingerboard. Now pass the string through the slots between the E and A string posts for the bottom E string, and the E and the B string posts for the top E string.

Now loop the string around behind the string where it entered the post originally so that when you turn the machine heads the string clamps onto itself in a simple locking manoeuvre. Cut off the excess string and tune to pitch. Repeat this procedure for the A and B strings, but passing them this time through the slots between the A and D strings (for the A string itself) and the B and G string (for the B string itself).

Finally, repeat the procedure for the D and G strings, passing the strings through the slots between their relative posts and the tops of the slots. This method ensures that the guitar reaches pitch more quickly and stays in tune for longer.

re-stringing classical guitars

When stringing a classical guitar, you must first tie the strings to the bridge correctly. To do this, pass the string through the hole behind the saddle and pull it through until you have approximately 6-8cm of string visible. Now pass the string back over the top of the bridge and loop it under the string where it passed through the hole. This will create a loop where the string passes over the bridge. Take the remaining few centimetres and pass it two or three times through the loop so that, when you pull on the main length of the string, it tightens the loops into place.

Now use the following method to attach the strings to the machine heads. Starting with the two E strings, pass the strings through the holes in the metal machine-head posts, pulling them through so that only a small amount of slack is evident over the fingerboard.

Now pass the strings through the slots between the E and A string posts for the bottom E string and the E and the B string posts for the top E string. Then loop the string round behind the string where it entered the post originally so that, when you turn the machine heads, the string clamps onto itself in a simple locking manoeuvre. Cut off the excess string and tune to pitch.

Repeat this procedure for the A and B strings, but this time pass them through the slots between the A and D string (for the A string itself) and the B and G string (for the B string itself). Finally, repeat the procedure for the D and G strings, passing the strings through the slots between their relative posts and the tops of the slots.

The benefits of this self-locking technique are that the strings reach concert pitch quicker and stay in tune considerably longer, thus remedying a particular problem for classical guitars.

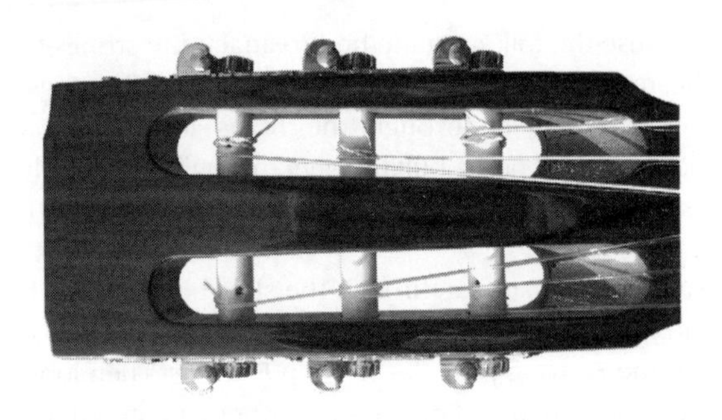

Classical guitars can slip out of tune if not strung correctly

re-stringing electrics

Electric guitars have many different sorts of machine heads fitted. Some come with standard machine heads, others come with locking machine heads. Some guitars are fitted with a locking nut which, as the name suggests, clamps the strings at the nut. This style of guitar also has a special tremolo system fitted, where the strings are locked into place just behind the saddle. This system involves cutting the ball end off the string and then clamping the string into place with an Allen-key clamping system.

Guitars fitted with locking machine heads are generally fitted with tremolo systems such as those found on Stratocasters. There are various forms of these machine heads on the market, but most function via a knurled nut on the underside of the head which comes into contact with a pin that passes through the string post and clamps the string into place. Guitars fitted with this style of machine head are very easy to re-string – you simply pass the string through the eye of the post, pull the string tight to remove the slack and then tighten the knurled nut.

In my opinion, it's just as easy to re-string a guitar with regular machine heads and, provided that the strings are attached properly, it'll offer the same tuning stability as locking heads. Again, you simply pass each string through the eye of the string post, then pass the excess string around the post and under the string where it originally entered, then pull it up and back over so that, when you tighten the machine head, the string locks onto itself.

Finally, whenever you change your strings, it's a good idea to lubricate the nut slots with a HB pencil. As we saw earlier, this increases tuning stability and helps prevent the strings from sticking.

tremolo units

Most tremolo systems work on the principle of counteracting the string tension with springs. By changing the tension of the springs, you can alter the angle at which the bridge plate "floats", although this only applies to the conventional Strat tremolo and the Floyd Rose design.

The Kahler-style tremolo has an Allen-key adjustment point on the top of the assembly so that the angle of the tremolo arm can be changed. (It's always best to refer to the manufacturer's instructions when adjusting this type of tremolo.)

Traditional Stratocaster tremolo systems should be set so that the arm can be pulled up as well as pushed down, thus allowing the pitch to be increased as well as decreased. This involves setting a gap of around $^{3}/_{8}$ of an inch between the back of the tremolo plate and the body of the guitar. Two screws attach the spring plate to the guitar body.

The bridge plates on Floyd Rose-style tremolo systems should be set so that they're parallel to the guitar body. This is achieved by adjusting the spring tension with the two large screws securing the spring claw in the back of the guitar. Tightening the screws brings the back of the bridge plate closer to the body, while slackening them moves the bridge plate further away.

The same basic principle applies to the conventional Strat bridge. When adjusting the spring tension, do so in small increments and retune to concert pitch after each adjustment. This will maintain the correct balance between the two forces involved and make setting the tremolo a much easier job.

After making these adjustments, check your action height and intonation again, as they may have changed during this procedure.

pick-up height

The correct height setting for the pick-ups is a compromise between obtaining the maximum signal strength from the strings without incurring the unwanted "false tones" caused by the pick-up magnets being too close to the strings. The pick-ups should be set in the following manner.

Firstly, hold down the strings at the last fret. Measurements should be taken from the top and bottom E strings to the top of the pole pieces. For humbucking pick-ups, this measurement should be approximately $^1/_8$ of an inch on the bass side and $^1/_{16}$ of an inch on the treble. For single-coil pick-ups, such as those found on Strats and similar guitars, it should be $^3/_{16}$ of an inch on the bass side and $^1/_{16}$ on the treble.

For bass guitars set the height to $^1/_4$ of an inch on the bottom E and $^1/_8$ of an inch on the G string.

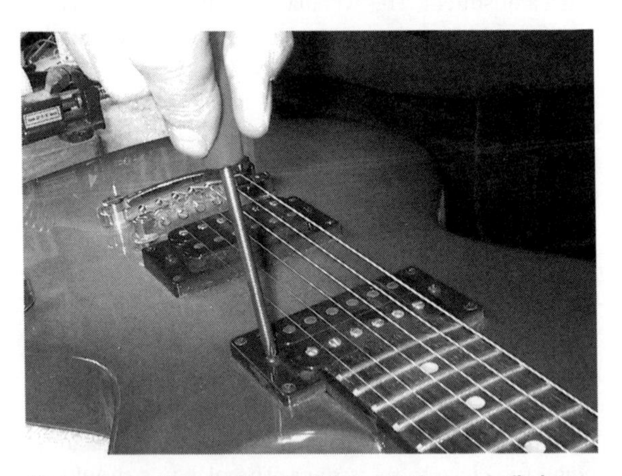

Use a ruler to verify the height of the pick-up pole pieces under the bass and treble E strings and adjust if necessary

troubleshooting: pick-ups, pots and switches

Electrical problems are quite common on guitars. Faulty switches and noisy pots are the bane of most guitarists' lives. If you don't want to experience these kinds of problems during a gig, your best course of action is to maintain your instrument regularly. This is quite simple to do, and in reality all you need is a few basic tools.

Generally, it's just a case of dirty contacts causing the problem rather than faulty soldering or poor wiring. It pays to clean your leads regularly with an electronic contact cleaner. There are various products on the market and most high-street electrical or electronics suppliers stock such items.

Even the internal components of your guitar should be cleaned at least once a year. You should use a lubricating electronic cleaner for your guitar's pots and switches. Just spray a small amount into the pots, taking care not to get any onto the guitar's finish. Pots can wear out, like most other things, but if the pot becomes extremely noisy or scratchy and a squirt of contact cleaner doesn't resolve the problem, you'll probably have to fit a new one.

The same goes for toggle switches: these can often become noisy or cease to function correctly. Getting hold of spare parts for guitars is a lot easier than it was several years ago; most major manufacturers carry spares and some even let you buy directly from them by credit card.

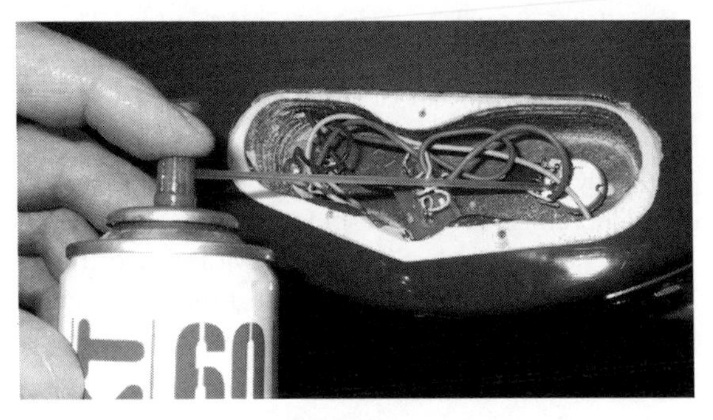

A small spray of contact cleaner will clean out dirty pots

Before you start any major work on your instrument, you should make a careful diagram of the wiring layout. (We'll be looking at soldering techniques a little later, so don't worry if you haven't done it before.) If the guitar ceases to function completely or a pick-up stops working, you should trace the fault with an electrical multimeter. You may need to disconnect or unsolder the suspect component to perform a simple resistance test.

You should also check your guitar lead and amplifier for faults. Fault tracing can be quite complex, and if you're inexperienced in this field, this job is probably best left to a professional repairer. However, if you enjoy dabbling with a soldering iron and have some experience with electronic circuitry, you should find that performing a continuity test is relatively simple. Sometimes, the problem can be due to a short circuit, which can often be found only by using a multimeter.

In truth, the electrical circuit of a guitar is relatively simple, but if you've never experimented with the internal workings of an instrument before, it can be quite off-putting. Assuming that you have some experience with the workings of guitar electronics, you should test the pick-ups first.

The impedance of most humbuckers ranges between 8kohms and 16kohms, whereas the impedance of single-coil pick-ups is between 4kohms and 8kohms. You should then check the functions of the pots and the switches. Look very carefully at the contacts on the switches and pots to see whether there are any stray whiskers of wire that could cause a short circuit.

Now check for continuity between the switch and the jack socket. Replace any faulty components that you find and check the wiring for any dry solder joints, which are dull grey in appearance whereas regular solder joints are bright silver. Dry joints can cause problems because they don't conduct efficiently.

acoustic guitar pick-ups

Electro-acoustic guitars are often fitted with pre-amplifiers powered by batteries. This is the first thing to check if a fault should arise (typically distortion) or if the guitar ceases to function. Even if you don't play regularly, you should change the battery at least once a year in case it starts to leak. Other common problems that might arise include poor string-volume balance, hum and popping or crackling noises.

Poor string-volume balance can be caused by a faulty pick-up: sometimes one of the piezo elements inside the pick-up ceases to function so that it can no longer convert string vibrations into electricity. Unfortunately, there's nothing that can be done about this, and you'll need to get a new one. Another cause for low string volume is a poorly fitted bridge saddle. You should make sure that the base of the bridge saddle is absolutely flat so that it comes into perfect contact with the piezo elements. This can be done by sanding the base of the saddle with a piece of fine sandpaper.

Place the sandpaper on a flat surface and gently rub the saddle backwards and forwards while keeping it in an absolutely vertical position. Now re-install the saddle, tune the guitar to pitch and try it again.

The remaining issues are more problematic. Any electrical noise could be due to a number of problems. It could be dirty contacts, faulty wiring, a bad earth connection on the pick-up or faulty elements inside it. All of these will have to be checked out and the problem found by a process of elimination. If a fault isn't obvious, always consult a professional repairer.

soldering techniques

Once you've mastered a few basic techniques, soldering is actually quite simple. You should always keep the tip of the iron clean and "tin" it regularly by applying fresh solder to it, especially before you start soldering. Also, you should practise soldering on a few scraps of old wire before you tackle the real thing.

Equipment-wise, you'll need a 25W iron with a range of tips for general-purpose soldering, and a soldering gun (50-60W) for heavy-duty work. For electronic work, it's advisable to use a multicore solder, which has a resin flux contained within the solder itself. There are various gauges available, but it's useful to have at least two different gauges, one light gauge for fine work and a heavier gauge for soldering braided pick-ups and other heavy-duty components onto the chassis of control pots.

To achieve a good joint, apply the solder at the same time as you apply the iron. This will avoid a dull and grey-looking dry joint.

In the process of soldering, be careful not to melt or burn any of the surrounding cables or wiring – and this includes your fingers! Don't use too much solder, either, as cleaning it up afterwards can be a tricky business. As with everything, practice makes perfect, and this is especially true if you wish to repair faulty joints rather than create more problems due to poor technique!

essential gig toolkit

Take a few basic tools around with you even if you don't gig on a regular basis. Your toolkit can be as simple as a couple of screwdrivers (cross- and flat-head), a pair of pliers, wire cutters, a soldering iron, solder, a torch and electronic contact cleaner. If you're using any battery-operated equipment, keep spare batteries with you.

You should also carry a spare lead at all times. If you have a valve amp, take some spare valves as well. You don't need a full set, just one or two of the power valves and one of the small valves used in the pre-amp section. It's a good idea to take spare fuses, too, and you can find the rating of the fuses on the back of your amplifier. Finally, always take a spare set of strings!

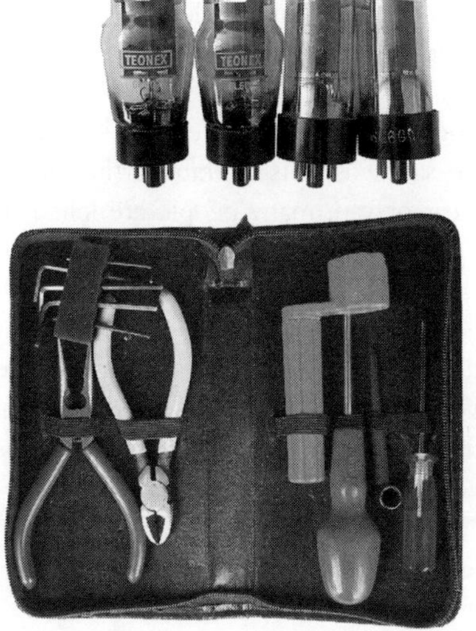

A few basic tools and spare parts will enable you to tackle minor technical problems at gigs or rehearsals

chapter 3

amplifiers and combos

Most amplifiers and combos lead a tough life – unless, of course, you never take your pride and joy out on the road! They tend to spend their lives in smoky pubs and in the backs of vehicles and are often stored in damp garages.

Before we get into this chapter in detail, it's wise to consider first the safety aspects of transporting heavy equipment in a car or van; many people are killed or injured each year through not taking suitable precautions when travelling in vehicles loaded with inadequately secured heavy goods. If you think that your equipment may have suffered due to bad storage conditions, have it checked over by a professional repairer. Damp environments can cause more problems than you might expect. Here are a few dos and don'ts:

- **Don't** leave your equipment unattended – a thief can strike at any time or place;

- **Don't** use guitar leads as speaker leads – they can actually damage your amplifier;
- **Don't** use speaker leads as guitar leads – they don't cause damage, but the buzz will be unbearable;
- **Do** have your equipment checked regularly for electrical faults;
- **Do** keep your equipment in a dry place;
- **Do** check all connection cables on a regular basis for faults and replace as necessary;
- **Do** purchase suitable waterproof covers for your amplifier and speaker cabinets.

pedals and rack-mounted effects

Over recent years, multi-effects units have become increasingly popular. It's easy to understand why this is the case: virtually everything you need to produce different guitar tones and effects comes in one box. Even the low-priced units are capable of producing great sounds and realistic amp tones.

The key to getting good sounds out of these units is, I'm afraid to say, reading the manual. As much as we all hate this procedure, it's the only way to get the most out of these devices. Take your time in getting to know the basic settings and sounds before turning to their digital effects and all of the commands necessary to edit and program them.

Don't be tempted to try to create 128 patches of killer sounds within five minutes of getting your new piece of kit out of the box – it just ain't gonna happen! Most serious guitarists work hard on creating maybe half a dozen patches of which two or three may be really good, clean sounds and the rest are crunch and overdrive for rhythm and lead work. Try to recreate the sound as accurately as you can according to the particular amplifier you have in mind. After this, you

Rack-mounted effects such as Line6's POD Pro come with a huge range of built-in effects and amp tones

can add in all your favourite reverbs, delays and choruses knowing that you're starting with a good basic sound.

Foot pedals and pedal-effects units probably receive the most wear and tear on the road, and these should be regularly inspected and cleaned externally, and their jack sockets should be squirted with electronic cleaner in order to avoid problems with the signal path.

Providing you take care of your equipment, you can expect it to perform properly for years. This cleaning routine isn't something you need to do every time you go out, but try to keep a little time aside to conduct some basic maintenance at least on a monthly basis.

speakers and cabinets

Speakers and cabinets tend to look after themselves, but it's a good idea to check them over every now and again. Quite often, jack sockets can become damaged or dirty and may require repairing or cleaning. Speakers may occasionally fail without you noticing anything unusual in a gig situation, so you should check them occasionally by taking the back off the speaker cabinet and checking each speaker individually both visually and with a multimeter.

Most guitar loudspeakers are either 8ohms or 16ohms. Anything radically outside this value indicates a fault and the speaker should be repaired or replaced.

Never mix speakers of different impedance or power rating in the same cabinet. If you do this, it may cause one or any of the speakers to fail – and it could even damage the amplifier if the impedance isn't right.

You'll find impedance ratings labelled on both your amp and speaker cabs

If you have a valve amplifier, it's absolutely essential that the impedance of the speaker cab correctly matches the impedance setting on the amp. There's generally an impedance switch on the back panel or dedicated jack sockets with an impedance value next to them. Valve amps can suffer serious damage due to impedance mismatching.

Transistor amplifiers are a bit more forgiving – as long as you don't go any lower than the recommended minimum impedance, almost any speaker cabinet will work. The one thing that transistor amplifiers don't like is short circuits. Although many newer amplifiers have short-circuit protection, you should still take great care if you're making your own speaker leads.

If you do decide to make your own speaker leads, ensure that the cable is heavy-duty enough to take the power handling of the amplifier. Mains cable should

Speakers and cabinets are robust but will benefit from routine maintenance

suffice, as it's robust enough for life on the road and easily obtainable. I suggest using at least five-amp cable to be on the safe side. Probably the best stuff to use is the bright orange mains cable used for power mowers.

maintenance

With regard to major electronic or electrical repair, there probably isn't much that you can do without jeopardising your safety. However, you can perform visual inspections on a regular basis and replace valves, fuses and perhaps even the odd component or two, should problems occur.

Transistor amplifiers and most rack-mounted multi-effects units are what they term "not user-serviceable". This, in effect, means that there is little or nothing you can do if the unit goes wrong. Of course, this doesn't mean that there aren't things that require attention – on the contrary, you should check regularly that the units are securely mounted in their flightcases and that leads and peripherals are in good condition.

troubleshooting

If your amplifier develops a fault during a gig, switch it off immediately. Sod's law, of course, dictates that this will happen in the middle of a song, but nevertheless you should stop playing and announce that you need to take a break due to technical difficulties. If you're using a transistor amp, there's probably little that you can do, apart from checking fuses. Any major repair on these amps is best left to a professional technician.

The same is basically true with a valve amplifier. You can, of course, change the valves if they blow, but it's inadvisable to attempt any sort of major repair in a pub or on a poorly lit stage.

If you have suitable facilities at home, you could attempt to conduct some troubleshooting, provided that you're aware of the potential hazards involved. Most amplifiers – including solid-state versions – run at potentially lethal voltages, and once you remove the amplifier from its cabinet you're exposed to these dangers, so be extremely careful.

Before you make any connections to the amplifier, check that the external mains fuse hasn't blown. When testing an amplifier, even just to check whether the valves light up or not, you should have the speakers plugged in to avoid causing any damage to the amplifier's output circuitry – in particular the valves or the output transformer. If the amplifier is still dead or not functioning correctly, you should unplug all connections, find a suitable workspace and assemble your tools and test gear.

After switching off your amplifier, you should wait for at least ten minutes before removing the chassis. This is because the large smoothing capacitors in the amplifier circuitry can retain high voltages for several minutes, draining it slowly over that time.

The first thing to do is to test the internal fuses, and this can be done with your multimeter – most modern meters have a continuity setting, which is useful for such jobs. If the fuses are okay, make a visual inspection internally to see if anything has come adrift or a component has burnt out. You can usually smell burnt components, if there are any.

If you find a problem, such as a loose component or a broken wire, you can probably fix the fault there and then, but if you come across any faulty components, make a note of their electrical values (and/or part numbers) and ask your local dealer to source them from the relevant manufacturer or distributor.

Most distributors and manufacturers hold good stocks of spares, so you should have to wait only a few days for the new parts to arrive. In the meantime, you could remove the faulty part and check the associated circuitry for any faults that might have caused the component to fail in the first place. This may be something as simple as a short circuit or yet another failed component. When installing the new component, make sure that you apply fresh solder to the joint in order to avoid any dry joints. Also, make sure that you trim any excess wire from the soldered joint in order to avoid a potential short circuit.

test equipment and tools

The test equipment required to carry out basic amplifier servicing can be obtained from most high-street electrical dealers. You'll need an electrical multimeter, a 25W soldering iron (plus solder) and a selection of hand tools, including pliers, cutters and various screwdrivers.

With regard to multimeters, you don't require anything too sophisticated – something fairly basic that can measure volts, amps and ohms will suffice in most situations. You'll probably have some of the tools already, but should you need to buy one or two more, there are some good deals to be had at most DIY stores. Hunt around for bargains!

essential servicing

You should check your plugs and mains leads on a regular basis. Remove the top from each plug, check that the connections are tight and secure and look for any physical damage to the lead itself. If there's any damage at all to the lead, or if it operates intermittently, replace it at the first opportunity. Intermittent mains leads can seriously damage electrical equipment as well as cause frustration during gigs or rehearsals.

The same goes for any other lead that you use to interconnect various pieces of equipment. Intermittent jack-to-jack leads can also cause problems such as electrical instability (which is often inaudible but can cause severe damage to amplifiers) and loud pops and buzzes, which can cause damage to your speakers.

spares

It's always a good idea to take a few spares to gigs for your amp and effects units. This needn't comprise much more than spare fuses, an extra mains cable, spare batteries (if you use battery-powered effect pedals or small tuner), spare patch leads and a few spare valves if you have a valve amplifier. Most of what you need to carry with you to a gig has already been covered in the previous chapter.

storage

The best place to store your equipment is in a spare room or under the stairs, if the space is large enough. Never leave it in a vehicle overnight – this is asking for trouble and may also affect your car insurance.

If possible, don't store any equipment in a garage, where it may be vulnerable to burglars and the weather! Dampness is a real enemy of electrical equipment. However, if you have no other option, you can reduce some of the risk with a reasonable amount of thought and some suitable precautions.

First and foremost, fit a heavy-duty padlock to the door of the garage and possibly some sort of simple battery-operated alarm – if it doesn't wake you, it might wake your neighbours, and if nothing else, it's a deterrent. To help keep out the elements, you could construct a wooden plinth to raise the equipment above floor level. Again, though, it's difficult to get insurance on equipment that's stored in garages and lock-ups.

cases and covers

To help keep your gear in good condition and prevent damage caused by damp or adverse weather, you should invest in a set of flightcases. Alternatively, a set of padded waterproof covers is the next best bet. You should also consider a rack case for your multi-effects units and perhaps a small, lightweight case for your effects pedals. This will go a long way towards keeping your equipment in showroom condition, as well as protecting it from life on the road.

safety

Considering the high voltages that are present in guitar amplification, it's unwise to attempt repairs unless you have considerable knowledge of the subject. Even if you've had some experience with electronic repairs, a simple – yet golden – rule when working inside an amplifier or touching the internal workings of any electrical device is to keep one hand in your pocket at all times so that electricity can't travel through your body and electrocute you. It's also a good idea to have a friend or neighbour with you when you're doing this sort of work so that, if you do have an accident, there'll be somebody there to help.

Working on guitars isn't quite as dangerous, yet there's always the potential for eye injury should a string break when you're sighting the neck. It's a good idea to have a pair of safety glasses in your toolkit – you never know when you might need them.

When gigging, use a plugboard with a built-in RCD (Residual Current Device) for the backline and PA, unless the power consumption of your equipment exceeds 13A, in which case you may need extra boards. Try to make sure that these are connected to sockets that are on the same electrical phase (don't use extensions from different parts of the building). This will protect you from the hazards of faulty equipment and the associated risk of electric shock. If you use your own lights, don't plug them into the same plugboards that you're using for the backline and PA, as you might draw too much current and blow a fuse.

Finally, never stand a can of beer – or any liquid, for that matter – on your amp, cabinets or PA; one day it will fall off and possibly cause severe damage to the gear, you or both!

chapter 4

PA equipment

Choosing a PA system is probably one of the most difficult things that you'll have to do. It's difficult to get it right and not end up with something that's inappropriate for the situation. If you're buying a PA system collectively as a band, you need to decide on what your objectives and priorities are. There are very few bands that can afford large concert systems, and to be honest you probably don't need one, especially if you're gigging just for fun and solely in local clubs and pubs.

A typical pub/club system might comprise a 600W mixer amplifier combined with two full-range speaker cabs. This will provide enough power for the average band and enable you to mic up an electro-acoustic guitar or two and cope adequately with up to four vocal microphones. However, this isn't really big enough to fully mic up the drums – for that you need to go up a notch in power and probably fork out for a bigger mixer with more facilities and channels.

The style of music that you play may also have a bearing on the optimum PA size that you should aim for. For example, if you play in a folk-rock or acoustic band, your power requirements may be lower than if you play in a classic-rock or a heavy-metal band. Ironically, acoustic bands often need as much power at their disposal as some of the wattage-hungry rock bands, due to the problems of miking up acoustic instruments, but they don't generally require bass bins or larger speaker systems.

As a rule of thumb, if you add up the power of the backline and add an extra 100W for the drum kit, this will give you roughly the minimum power required for your PA system. For most bands, this will be somewhere in the region of 500W. If you play regularly in larger clubs or venues, then you'll probably require a multichannel mixer, power amps and monitors, along with separate bass bins and mid/hi speaker systems.

A decent monitoring system may need to be as powerful as the main PA. This is to provide you with enough "power headroom" to reduce the likelihood of feedback. For smaller bands and duos, it's possible to purchase monitor cabinets with built-in amplifiers, which are more cost-effective and take up less room.

If you play in a small duo and only occasionally use a machine for backing tracks, then one of the compact PA systems that are available may be suitable. These units range somewhere between 150-300W in power output and come with small, full-range speaker cabinets that are stand-mounted. Duos that require a bigger sound often need a PA system as big as that used by a regular band.

troubleshooting

Assuming that you've got a suitable PA system at your disposal, one of the biggest problems you're going to encounter is deciding where to site the PA amp and the speakers to get the most effective spread of sound. This, of course, will change from venue to venue, depending on the amount of room available, as well as the shape of the room. If you're lucky enough to have a friend whose ears you can trust – and who is prepared to double as a sound engineer – you could site your PA amplifier or mixer off-stage. One drawback to this is that you may have to buy a multicore cable and a stage box (to plug in the mics and other equipment) so that you can get a reasonable distance from the stage to enable him to mix adequately. If you have to do the mixing on-stage, one of you should assume responsibility and mix the rest of the band by running backwards and forwards between the room and the stage (while the band is playing) and adjusting the PA to achieve a good balance.

Where possible, the PA speakers should be sited in front of the band in order to avoid feedback. If this isn't possible, and if their instruments or microphones present a problem, you may have to move the various bandmembers into different positions. Be careful if you're walking in front of a PA speaker with a mic as this increases the likelihood of feedback occurring.

If the PA system doesn't function correctly, you should have a mental checklist of problems to look for. These include such things as:

- Is there power to the wall socket?
- If you have an RCD plug fitted, has it tripped?
- Is the system turned on?
- Is the master volume turned up?
- Is the channel volume turned up?
- Do you have a faulty mic or instrument?
- Is everything plugged in properly?
- Are any of the leads damaged or faulty?

99 per cent of the time, one of the above problems will be responsible for causing the system to fail.

mixers and power amps

There is such a wide range of PA equipment available today that you really need to seek specialist advice, according to the type of band and the style of music you play. Fortunately, there are many specialist dealers who can help you with your choice of equipment and help you decide whether you should opt for a multichannel mixer and separate power amp or a combined mixer amp.

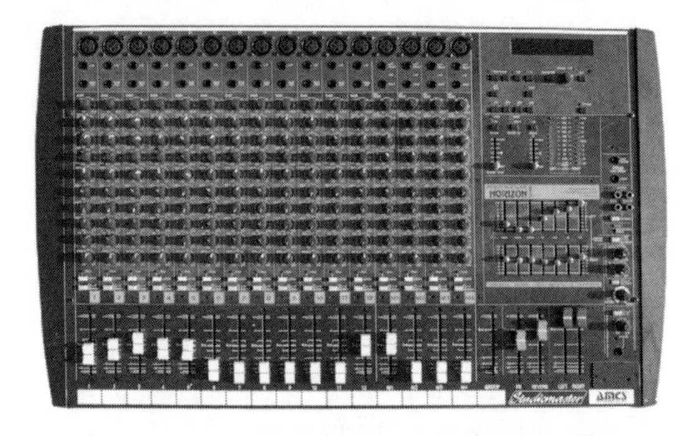

Choose a mixer that suits the needs of your band

If you play in a band that gigs only occasionally, you'll probably find that a combined mixer amp and a small pair of full-range speakers will suffice. Many such systems are available now at prices significantly lower than they were even just a few years ago. The slightly higher-priced versions also include such features as digital reverbs and delays.

If you have the budget and a suitable mode of transportation, you may decide that you want to purchase a larger system which comprises a mixer, a 1,000W (or more) stereo power amp and larger PA speakers. Multichannel mixers are also available in many varieties. The professional varieties cost big bucks and are only really worth it if you've got money to burn! For the average-sized semi-pro band, a 16-channel mixer with four subgroups should suffice. Something of this size will enable you to mix drums and keyboards separately and still have enough channels left over for bass, vocals and acoustic guitars.

Bigger is not necessarily better – and that goes for amps, too!

Power amplifiers also come in many varieties and power ratings. Most of them are stereo and are capable of delivering 100W per channel or more. The most popular style of power amplifier delivers around 500W RMS per channel. You'll need to consider the type of PA loudspeaker you're going to use them with as well, though, as it's pointless having a large power amplifier if you're just using a small pair of full-range speakers.

A simple tip is to perhaps buy two power amplifiers (both with a slightly lower wattage rating) rather than a single high-power one. This is so that, if one should fail, you will at least be able to finish the gig with the other. Also, on smaller gigs, you could use the other power amplifier for monitors rather than having to lug around two racks of amplifiers.

Once you've decided on the system that suits your requirements, don't forget to buy some sort of cover or flightcase, as knobs and sliders are very delicate and vulnerable to being knocked and damaged.

PA and monitor speakers

Choosing suitable PA speakers can also be quite tricky. The smaller mixer-amp systems tend to come with suitably sized full-range speaker units, which are often included in the price. Speaker systems over 500W per channel sometimes come with separate bass bins and mid/hi units.

Some of the new modular systems are very good and can be quite useful if you're regularly playing venues of different sizes. On the smaller gigs, you can take out maybe just a pair of cabinets, and on the larger gigs you could take out the full rig.

Don't be fooled by the small sizes of these new units, as they can deliver an enormous amount of power. Modern high-tech plastics are often used for the cabinets of these small PA units, which makes them lighter and allows for specialist built-in features such as handles and flying points for installation work.

Monitor speakers are available in passive or active formats and some even have built-in power amps. If you play larger stages, you may need a fairly beefy monitor driven by at least 500W of power, while for pubs and clubs you'll find that the smaller monitors with built-in power amps will suffice.

When choosing PA speakers, make sure that they're able to handle the power of your amplification. Again, you might need to seek the advice of a specialist dealer.

microphones

Microphones are available in all shapes and sizes, although my advice would be always to choose the best that you can afford. There are, of course, some bargains to be had with some of the cheaper range of microphones, but by and large the more expensive ones give better sound quality and offer better feedback rejection.

Microphones are often manufactured for specific purposes and may not be suitable for all-purpose work, so check with your dealer before buying

The pencil-style condenser microphones are generally used for acoustic instruments such as guitars or pianos as well as overhead miking for drum kits.

Larger condenser microphones are normally found in recording studios and aren't really suitable for live PA work – apart from which, they are expensive.

Many companies make hand-held condenser mics for vocalists. When choosing a microphone for vocal work, make sure you have one that has a unidirectional cardioid polar pattern to help eliminate feedback.

The most popular form of microphone for live work is the moving-coil dynamic variety. There are versions of this style of microphone available for vocals, instruments and drums, and just about every microphone manufacturer has something suitable in its range. Many make them for specific purposes, such as miking the bass drum of a drum kit, for example.

Although most microphones look fairly robust, you should avoid throwing them around while doing your impersonation of Roger Daltrey. It's a good investment to get some sort of case in which to keep your microphones. There are various styles available, and some recent versions are made of lightweight plastic and can hold up to six mics or more.

mic stands

Probably the most useful microphone stand is the boom variety. These can be used for just about any musical situation you can imagine. Whether you're a vocalist or a drummer, a boom stand will be useful.

This is especially true if you're a guitarist or bass player, as a boom stand allows more clearance.

The upright stands are good for posing in front of and throwing around, but if you have an instrument slung around your neck, they tend to get in the way.

It's sensible to avoid microphone stands that have a plastic base, as these break more easily than those made from metal. Again, buy the best that you can afford.

Boom mic stands are suitable for pretty much all situations

DI boxes

A DI box is a simple interface between an instrument and a PA mixer into which you can plug all sorts of things, from electro-acoustic guitars, bass guitars and keyboards to guitar processors and speaker simulators. Some are passive, while others are battery-operated or phantom-powered from the mixing desk. They are available in mono and stereo formats, depending on your needs. (Most keyboard players – or guitarists using effects processors – will require the stereo version.)

The cheaper units comprise a specially wound transformer and a few switches and jack sockets, whereas the more expensive units feature active electronics. Even if you can't see the need for a DI box at present, it's always useful to have one to hand, as sooner or later the need for one will crop up.

transportation

Years ago, transporting a full PA system required at least a small van. Fortunately, most modern systems are quite compact, so many bands now don't bother with a van (the modern compact systems are easy to transport in even the smallest car). If you have a larger system, however, it's safer to transport it in a van rather than a car, and if you don't have a van you should consider hiring one. The logic behind this is that, if you're playing in places that demand a larger PA, the budget should be there to hire a vehicle.

storage

Modern, compact PA systems are much easier to store indoors than the bulky PA systems of yesteryear. If possible, you should store all of your equipment indoors, as this will protect it from the elements and theft. If you have to keep your PA in a garage or outhouse, make sure that you take security precautions. Also, you should make sure that the equipment can't get damp, as this could cause it to malfunction.

When your gear is stored in a van or car, try to keep the vehicle in a locked garage or compound. If this isn't possible, find a parking place near to your house or venue. Also, if you regularly store equipment in a vehicle, it's sensible to fit a security device.

safety

Provided that you take sensible precautions and maintain your equipment regularly, you shouldn't have any problems with safety. However, there are a few things that you should be aware of, one of the most important being that you should not take your power source from a different electrical phase to that of the backline equipment as, under certain conditions, you may get buzz or hum from the equipment or, in the worst-case scenario, an electric shock!

If your PA doesn't draw too much power, you can probably source the mains supply from one socket (via a plugboard) to provide power for your backline as well as your PA. If your equipment draws more than 13 amps, take your power from additional sockets in the same room – not from any other room, unless you're sure that it's on the same electrical phase as the supply you're using.

If beer – or any liquid – is spilled over the PA or any of the backline equipment, don't touch it; switch off the gear, unplug it at once and don't use it again in the same evening, as it will compromise the safety of the band. Wait until the equipment has thoroughly dried out before turning it on and testing it. It's sensible to have it checked over by a technician before you use it again, as beer often leaves sugary residues that can cause short circuits. Finally, it's a good idea to fit an RCD (Residual Current Device) to your plugboard, in order to prevent electric shocks, and to check all your equipment and leads on a monthly basis. Get into the safety habit!

drum maintenance

Unlike guitarists, drummers don't usually like to be seen with a battered or filthy drum kit. To keep your kit functional and clean, it's a good idea to give it the occasional wipe-down with a cotton cloth that's been soaked in warm water and a mild washing-up liquid like Fairy and then, making sure it's wrung out thoroughly, give each drum the once-over. The drum heads can also be cleaned like this, but make sure you don't slop the water around and get the drum too wet! If you use textured or real hide skins, be careful not to slosh water all over them – just use a slightly damp cloth. A little bit of leather cleaner will work on hide skins that have a treated or polished finish. Dry the drums off carefully and then polish them with a wax-free polish such as Sparkle or Mr Sheen.

The drum and cymbal hardware should also be cleaned and lubricated. Use a wax-free polish to clean it and either WD40 or Three-In-One oil to lubricate moving parts, clamps, nuts and bolts. To remove rust or tarnish, try rubbing the affected areas lightly with a metal polish such as Brasso. Don't forget to take particularly good care of your drum pedals, as a lack of regular cleaning and lubrication can cause mechanical problems that may affect playability – which won't go down too well with the rest of the band. Again, a drop of oil will work wonders. Hardware should be cleaned and lubricated on a monthly basis to maintain its functionality.

cymbals

You may wonder why you have to clean your cymbals, but if you want them to stay sounding crisp and clear, the occasional wipe-over will work wonders. Modern cymbals are coated with a lacquer that prevents them from tarnishing, but unfortunately this coating doesn't prevent them from getting dirty! Marks from your sticks and the build-up of grime and grease from your fingers can make them look decidedly unattractive, even over a period of just a few weeks. In severe cases, it's a good idea to wipe them down using a cotton cloth and a mixture of warm water and Fairy, or something similar. When you've cleaned them satisfactorily, they should be polished with a wax-free polish such as Sparkle or Mr Sheen, which will help prevent the build-up of grime and keep them sounding great!

changing drum heads

After a few weeks of hard use, you'll probably notice that the sound and tone of your drum heads or skins begins to deteriorate. When it becomes really apparent or annoying, you should consider changing them. Professional drummers can replace their heads on an almost daily basis, but for the rest of us changing them a couple of times a year should suffice.

Once you've removed the old skin, you should check the drum internally for cracks or splits in the shell, and any problems here should be rectified before fitting a new skin. Fit the new skin in position and replace the hoop. Now adjust the lugs by hand until they are finger-tight and press downwards in the middle of the skin with the palm of your hand to take up the slack – re-tighten with your fingers and repeat the procedure. Then, using the key, tighten each lug by half a turn in a clockwise direction. Continue until you reach the required pitch. This method will ensure that the drum heads tune evenly, minimising "wolf tones" and ringing.

risers and rugs

If you play regularly, you'll know the problems associated with drum kits and cymbal stands sliding and moving around. A simple solution to this is to use some foam- or rubber-backed carpet that's just large enough to accommodate your drum kit. A piece measuring 2m by 1m should suffice. In the interests of getting return gigs, you should never nail your kit to a wooden floor or stage.

If you're handy at DIY, you could also consider building yourself a stage riser so that you can be seen above the rest of the band. Obviously, you'll need a mode of transportation big enough to carry this, so if you're lucky enough to have something as large as a Transit van you should easily manage to fit in a stage riser. If not, you may find that certain venues have them – most theatres and larger venues do. It always pays to ask!

To build your own, first make a folding wooden frame onto which you can place a piece of MDF board. Glue a piece of carpet to this board to prevent the kit from moving around. The riser needs to be only around 30-40cm high, and with a bit of thought you should be able to knock something up without breaking the bank.

cases

There is a vast range of drum cases available nowadays, so deciding on the correct one for your kit will basically depend on where and how often you gig, as well as your method of transportation. Although drums may look robust, the truth is that they are actually quite fragile and can be damaged easily. On many occasions, I've seen drummers wrap up their kits in blankets and put them in the backs of their cars! This is fine if you're not gigging regularly and just need to transport your kit to a new location, but if you're trying to put off buying cases, it's false economy in the long run. You should also consider the fact that, if you want to sell your kit at a later date, you may get less for it if it's damaged.

fibre cases

The traditional fibre drum case has been with us for many years and is still a favourite with many drummers. In fact, this is probably the world's most-sold style of drum case. As general workhorses, these cases are perfectly adequate for most purposes, and over the years drummers have relied on them heavily to protect their drums and other hardware.

These cases are made in two parts from compressed man-made fibre and resin, one half of the case fitting over the other to enable them to accommodate drums of varying heights. The two halves are generally held together by specially designed straps and buckles. Although not as sturdy as the more expensive flightcases, fibre cases are good value for money and are suitable for all circumstances except professional touring and carriage in the holds of aircraft.

soft bags

A recent drum-world development is the padded drum bag, similar in style to the guitar gig bag. These bags are made from a tough man-made material that resists tearing and splitting and is also resistant to light showers of rain.

Zip fasteners are fitted to the fabric lids of these cases, enabling easy access without the use of fussy belts and buckles. Because these bags are both lightweight and practical, they are becoming increasingly popular and will probably take over the fibre-case market ultimately. They offer reasonable protection from the bumps and thumps that most drummers subject their kits to, with the added benefit of folding flat once the drums are removed, thus taking up less room.

flightcases

These large, bulky items are the ultimate in transporting drum kits – if you have the means of transportation, that is! There are several major manufacturers of flightcases in the UK alone, but rarely do you see their products for sale in shops, although good dealers can get them for you made to order. Most manufacturers are prepared to deal with customers directly, as this makes it easier for them to obtain the required information on kit size and number of drums.

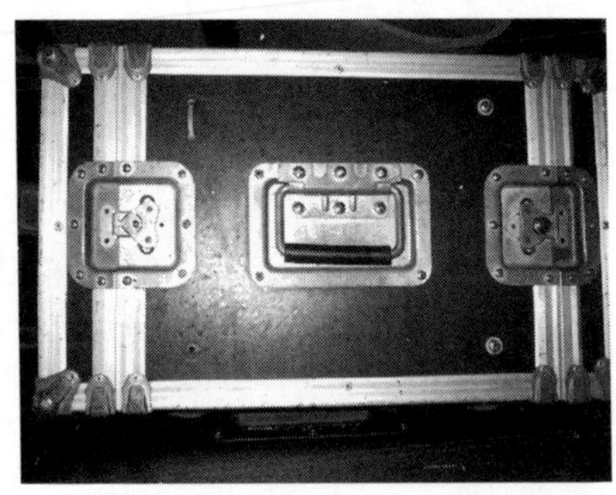

A heavy-duty case will protect your kit from the rigours of life on the road

A heavy-duty flightcase is normally made from high-quality plywood covered in a protective metal sheet or hard-wearing plastic. The edges of the case are protected by aluminium corners and heavy-duty latches and locks are fitted for security. The insides are usually well padded to protect the drums and hardware. The larger professional versions normally house the whole kit – including cymbal stands – in separate compartments.

The cost of professional flightcases is prohibitive for many players, and only the lucky few who have the money, transportation and – yes – roadies can afford the luxury. However, new developments in lightweight flightcases are taking place all the time, so keep a lookout and network with other drummers and dealers for information.

transportation

One thing's for sure: you can't go by bus! Transporting drums involves a vehicle at least the size of a family

A partition between you and your kit is a sensible safety precaution

estate car. I've known drummers trying to cram their kit into a small car and unwittingly compromising their safety in the process.

If you can afford a van, this is the sensible option, as you'll be able to put up a partition behind the driver and passenger seats for protection and safety.

If the whole band has to share a single vehicle, there may still be space problems preventing you from taking your whole kit and riser (if you've got one) to a gig. It's also essential to make sure that all drivers are covered by insurance and that you let the insurance company know that you're transporting expensive musical equipment. Some insurance companies simply won't deal with this sort of risk, especially when it comes to bands with younger members, so you may have to shop around. Contact your local branch of the Musicians' Union for more advice.

storage

If you have to store your kit for a longer period of time than the usual couple of weeks between gigs and rehearsals (ie perhaps a month or more), it's sensible to take a few simple precautions to ensure that it doesn't rot away in the meantime! Suitable places for long-term storage are under the stairs, in a spare room and in a dry outhouse. Unsuitable places for long-term storage are in a shed or garage (unless you have absolutely no choice in this), in an attic (unless it's well insulated) and in a vehicle.

Clean, polish and lubricate your kit before storing it away, or it could suffer badly from the effects of humidity and adverse weather conditions – especially if you have to store it in a shed or garage.

Place the drums in their cases along with a few small bags of silica gel to absorb any moisture or condensation that may build up. Don't cover them with blankets, though, as they may get damp, which could ultimately cause the hardware and fittings to rust and the shells to warp.

essential tools and spares

Believe it or not, even drum kits break down occasionally, and sooner or later it's going to happen to you! Although hardware is more reliable and robust nowadays, it's essential to be prepared in case something does go wrong. If you get into the habit of carrying with you a bag stocked with basic tools and spares, you should never come unstuck.

As already mentioned, you should perform basic maintenance on your kit on a monthly basis, as you can never predict when that head is going to split or the drum pedal will fail. You don't need a specialist set of tools to keep your kit alive; just the basics will suffice.

Smaller flightcases are ideal for carrying around spares and tools

Here is a suggested list of tools and spares to take with you to a gig:

- A drum key
- A pair of pliers
- An adjustable spanner
- A medium hammer
- Various screwdrivers
- A roll of gaffer tape
- A can of wax-free polish

- A can of WD40
- Three-In-One oil
- At least one spare pair of sticks
- A spare set of drum heads
- A length of stiff wire (for lashing and securing broken stands and pedals)
- Felt washers (for cymbal stands)
- An extra beater for the drum pedal

afterword

With the knowledge you've gleaned from the pages of this book, you should now be in a position to look after your kit, as well as that of your bandmates, and get a more productive, longer life from your instruments. You'll find that the simple techniques and instructions provided here will not only serve to increase your technical confidence, when dealing with problematic hardware, but that your playing experience will also be greatly enhanced. Guitarists and bass players for whom action height has never been an issue will find their instruments more satisfying to play once set correctly, and drummers will enjoy playing well-maintained, user-friendly kits. Finally, rehearse, gig and enjoy it!